Trick or Treat, Bugs to Eat

Words by
Tracy C. Gold

Pictures by
Nancy Leschnikoff

sourcebooks eXplore

Trick or treat!
Smell my feet!

Give me lots of
bugs to eat!

From our cave,
we're so brave,
we fly out in
one big wave.

Off I fly,
soar so high,

gliding through the
evening sky.

Hear my calls
bounce off walls,

echoing as
darkness falls.

Moths take flight
to the light.

I'll give them a
sudden fright!

Yummy flies
with big eyes!
What a trick-or-treat surprise!

Ooo, crickets' song.
Did I hear wrong?

No! I'll eat them
all night long.

CRUNCH

Beetles crunch,
beetles munch,
I'll have them for
midnight lunch.

Flap my wings!
Please, no stings!
When I eat wasps,
my heart sings!

I'm so fast,

whizzing past,

finally I'm
full at last.

What a flight,
lots of bites!

Halloween's my **favorite** night!

Many different types of bats live in the United States. Most of them eat bugs. Some bats eat fruit and help to pollinate plants. There are only three types of bats who drink blood from animals, but none of them live in the United States.

All bats have a tragus near their ear. It's a flap that helps them with echolocation. Bat wings have bones in them similar to the bones you have in your hands. Bats even have thumbs!

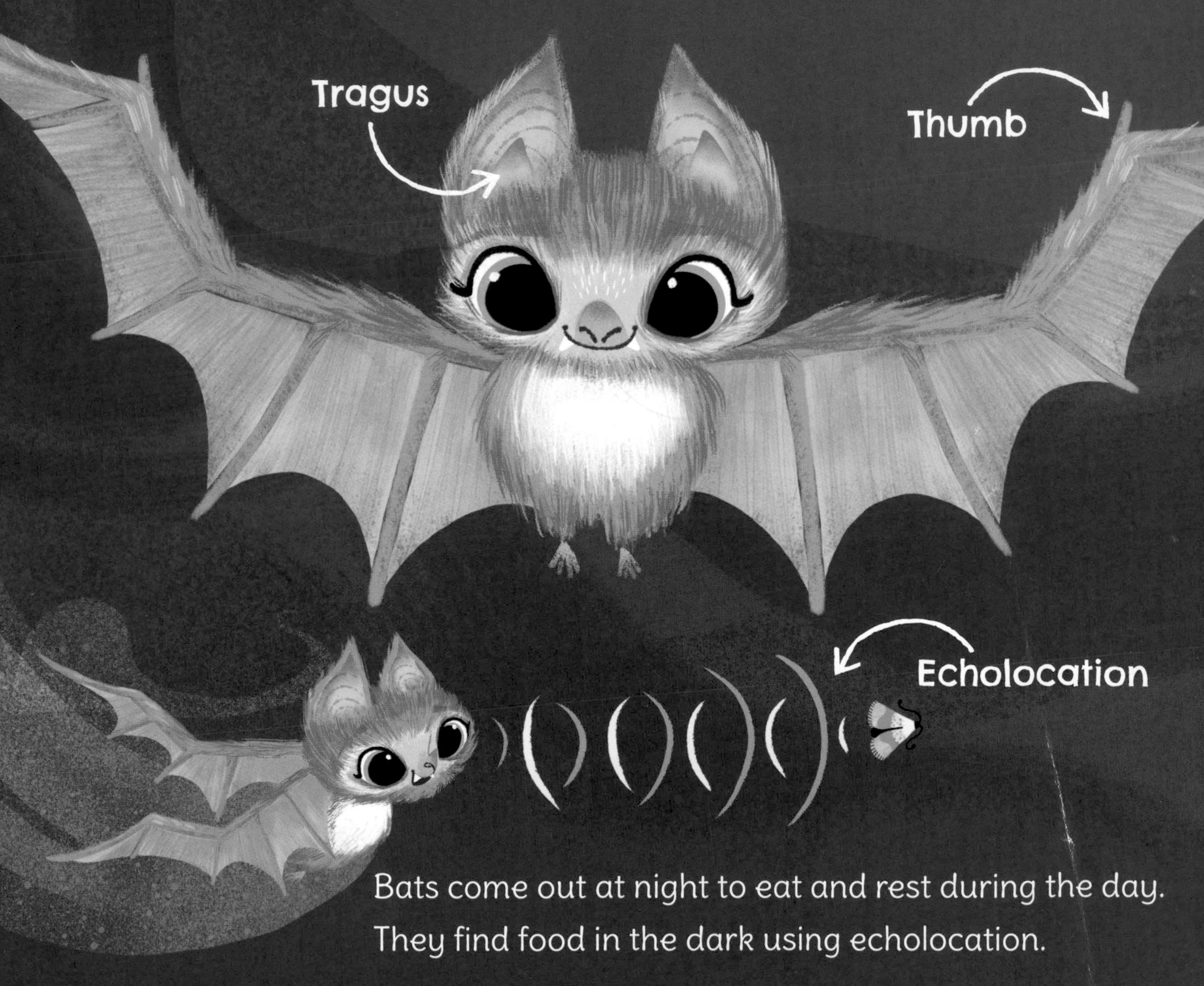

Bats come out at night to eat and rest during the day. They find food in the dark using echolocation.

Bats echolocate by making noises and waiting to hear the noise bounce back from objects—and insects—around them! Can you learn about your environment by listening to echoes? Try it out!

Bats live in caves and other dark places like the underside of a bridge—or maybe even your attic.

If you see a bat while you're **trick-or-treating**, don't be scared. Thank it for eating bugs so they don't bite you. Remember, if you see a wild animal, even a cute bat, don't touch it because you could hurt or scare it!